The Ransomware Threat Landscape

Prepare for, recognise and survive
ransomware attacks

The Ransomware Threat Landscape

Prepare for, recognise and survive ransomware attacks

ALAN CALDER

IT Governance Publishing

IT Governance Publishing Ltd
Unit 3, Clive Court
Bartholomew's Walk
Cambridgeshire Business Park
Ely, Cambridgeshire
CB7 4EA
United Kingdom
www.itgovernancepublishing.co.uk

First edition published in the United Kingdom in 2021 by IT Governance Publishing

ISBN 978-1-78778-278-5

ABOUT THE AUTHOR

Alan Calder founded IT Governance Limited in 2002 and began working full time for the company in 2007. He is now Group CEO of GRC International Group plc, the AIM-listed company that owns IT Governance Ltd. Prior to this, Alan had a number of roles including CEO of Business Link London City Partners from 1995 to 1998 (a government agency focused on helping growing businesses to develop), CEO of Focus Central London from 1998 to 2001 (a training and enterprise council), CEO of Wide Learning from 2001 to 2003 (a supplier of e-learning) and the Outsourced Training Company (2005). Alan was also chairman of CEME (a public private sector skills partnership) from 2006 to 2011.

Alan is an acknowledged international cyber security guru and a leading author on information security and IT governance issues. He has been involved in the development of a wide range of information security management training courses that have been accredited by the International Board for IT Governance Qualifications (IBITGQ). Alan has consulted for clients in the UK and abroad, and is a regular media commentator and speaker.

CONTENTS

INTRODUCTION

In May 2017, someone in the UK's National Health Service clicked a link in an email. Within hours, hospitals, surgeries and offices across the country were being shut out of their own systems, causing delays to critical healthcare. For several days, the NHS was paradoxically in a state of both paralysis and upheaval: crucial health information couldn't be accessed, appointments couldn't be made or checked, and patients were left waiting; meanwhile, IT teams scrambled to contain the outbreak and searched for a solution.

WannaCry, the ransomware to blame, went far beyond the NHS: it affected around a quarter of a million computers in 150 countries, including devices used by critical national infrastructure, industrial control, state agencies, multinationals and private citizens alike. The final bill was in excess of US$4 billion, with the attackers raking in nearly $400,000.[1]

None of this should be news today. the ransomware attack made headlines across the globe and probably did more to get companies to update their software and operating systems than any other single action in history. However,

---

[1] It's worth noting that this is a relatively small take for the criminals. The REvil ransomware developers claim to have made $100 million in a year, and want to make $2 billion. Ionut Ilascu, "REvil ransomware gang claims over $100 million profit in a year", *BleepingComputer*, October 2020, *www.bleepingcomputer.com/news/security/revil-ransomware-gang-claims-over-100-million-profit-in-a-year/*.

as with many other crises, lessons are often only remembered for a short period. Once the drama has faded and other disasters fail to emerge, it's very easy to settle back into old habits.

While there was certainly a lot of luck involved in the WannaCry outbreak – both enabling the attack to be effective and to reduce its longevity – these elements of luck should really serve more as warnings:

1. The NHS was not the target. Like most malware, once it was released into the wild, the ransomware would simply cripple whatever systems it could whenever it was able to.
2. The 'kill switch' that was discovered let the world recover relatively quickly.
3. WannaCry targeted a vulnerability for which there was already a patch, so it could only affect systems with out-of-date components.

All of these can be read to work in the public's favour: it was random chance that it happened at all, it was all over with quickly, and it only affected out-of-date systems. Surely that means the whole thing was just dumb luck on the attackers' part? Sadly, there's a better way to interpret these details:

1. Like any other malware, it was untargeted. This means that a more effective, more damaging piece of malware could have been spread in the same way. Certainly, more targeted attacks have caused

significant damage albeit on a more limited geographical scale.

2. The overwhelming majority of ransomware doesn't have a kill switch that's so easy to discover.

3. WannaCry's impact demonstrated just how much of the world's computing resources were out of date. It's likely that a comparable proportion today is also out of date.

We also need to remember that malware of all kinds evolves: its developers want to have the best product out there, and they're in a constant arms race with software vendors that are trying to plug gaps in their code. The fact that no other ransomware has had the same global impact since isn't really a vindication of modern cyber security: it's just another matter of dumb luck.

No organisation – whether multinational, SME, micro, government body or charity – should rely on dumb luck for its security.

This is especially true with regard to ransomware, which is the fastest growing malware in the world. In 2015, it cost companies around the world US$325 million, which rose to $5 billion by 2017 and is set to hit $20 billion in 2021.[2] Logically, the only way it can achieve such staggering

---

[2] Cybersecurity Ventures, "Global Ransomware Damage Costs Predicted To Reach $20 Billion (USD) By 2021", October 2019, *https://cybersecurityventures.com/global-ransomware-damage-costs-predicted-to-reach-20-billion-usd-by-2021/*.

growth is by either increasing the number of targets hit or significantly increasing the damage of an attack.

The statistics currently support the latter, as there has been relatively low growth in the number of ransomware attacks. SonicWall reports that 2019 actually saw a 6% decline in the number of ransomware attacks[3] – but the damage was considerable: Emsisoft Malware Lab found that ransomware in 2019 cost the US alone more than $7.5 billion.[4] It's also worth remembering that a reduction or stalling in the number of ransomware attacks isn't proof that they are in decline – it's just as likely that there will be a resurgence as soon as a key vulnerability is identified.

This doesn't paint the full picture, however, as ransomware is only part of the overall malware biome. During the same time period, Malwarebytes observed a 73% increase in Emotet infections – Emotet is a Trojan often used to deliver ransomware payloads.[5] An increase in this sort of malware is often linked to an increase in ransomware. In this instance, it likely indicates a shift in the methods by which ransomware reaches its targets to a more sophisticated approach that's harder to combat.

---

[3] SonicWall, *2020 SonicWall Cyber Threat Report*, February 2020, *www.sonicwall.com/resources/2020-cyber-threat-report-pdf/*.

[4] Emsisoft, "The State of Ransomware in the US: Report and Statistics 2019", December 2019, *https://blog.emsisoft.com/en/34822/the-state-of-ransomware-in-the-us-report-and-statistics-2019/*.

[5] Malwarebytes Labs, *2020 State of Malware Report*, February 2020, *https://resources.malwarebytes.com/files/2020/02/2020_State-of-Malware-Report-1.pdf*.

This combination of factors makes ransomware especially concerning. An attack that's harder to detect and prevent, and causes considerably more damage is like the development of the intercontinental ballistic missile (ICBM) in the twentieth century: a device capable of striking anywhere in the world and causing catastrophic damage, with few countermeasures possible. ICBMs are expensive and don't usually point at countries that don't pose a threat, however, while ransomware attacks are cheap and anyone could be targeted.

It's clearly the duty of all business leaders to protect their organisations and the data they rely upon. This must extend to doing whatever is reasonably possible to mitigate the risk posed by ransomware.

This duty is backed by the law. An increasing body of legal and regulatory requirements establish that organisations must protect all kinds of data from threats including ransomware. The European Union's General Data Protection Regulation (GDPR), for instance, mandates that organisations implement "technical and organisational measures to ensure a level of security [for personal data] appropriate to the risk".[6] As the GDPR requires that personal data be accessible to the people it refers to, ransomware poses a clear risk. Some ransomware also exfiltrates data, which is a clear threat under the law. The Regulation binds every country within the EU and, through contracts, a vast swathe of organisations across the globe.

---

[6] GDPR, Article 32.

In the US, the California Consumer Privacy Act (CCPA) similarly gives consumers the right to seek legal redress if their data is interfered with by an unauthorised party.[7] While it does not specify loss of access to this data, it's worth remembering that ransomware doesn't necessarily just lock down information and systems – the same data is often sent to the criminals behind the attack, who can threaten to release the data to the public or onto the Dark Web if the ransom isn't paid.

It's not just about personal data, either. Comparable laws demand that organisations protect their systems from interference, such as under the EU's Directive on security of network and information systems (NIS Directive), which focuses on critical infrastructure and digital service providers. Under this law, EU member states are required to establish laws to minimise the impact of "harmful actions intended to damage or interrupt the operation of the systems", which clearly applies to ransomware.[8]

Beyond legal requirements, organisations around the world are bound by contractual obligations to protect their data and systems. The Payment Card Industry Data Security Standard (PCI DSS), for instance, is a set of requirements for organisations involved in card payments, and specifically requires that information and systems involved in these transactions be protected from threats such as ransomware and other malware.

---

[7] CCPA, s. 150.

[8] NIS Directive, Recital 2.

This extends to obligations to investors and other stakeholders to protect the organisation. A number of attacks focus on locking up and/or stealing intellectual property and threatening to release it. In addition to reputational damage, the organisation can easily suffer long-term harm from having its intellectual property (IP) released into the wild.

Furthermore, an increasing number of organisations demand that their suppliers provide guarantees of protection. This isn't just a sensible precaution: suppliers and other third parties are a key weakness for many organisations, and criminals know this. Target's infamous data breach in 2014 was brought about because a supplier had poor security (and it's fair to say that Target's processes were also inadequate, exacerbating the situation).[9]

Business leaders are burdened with a number of responsibilities. Ransomware poses a threat to those responsibilities on a number of fronts: aside from the simple, direct harm that it can cause, it also puts the organisation at risk of being discovered to not meet its legal and contractual obligations.

Cyber security isn't simple, though, and it's easy to be confused by the technology and the pace of changes. This book aims to cut through that and provide a clear pathway to protect your organisation, and to do so in a way that is

---

[9] For more information, visit:
www.theguardian.com/business/2014/jan/10/target-more-customers-data-stolen-breach.

easy to maintain and extend as the environment and your infrastructure evolves.

The following chapters set out in clear language how ransomware works to help business leaders better understand the strategic risks, then delves into measures that can be put in place to protect the organisation. These measures are structured such that any organisation can approach them. Those with more resources and more complex environments can build them into a comprehensive system to minimise risks, while smaller organisations can secure their operations with simpler, more straightforward implementation.

CHAPTER 1: ABOUT RANSOMWARE

In general, there are three types of ransomware:

1. Scareware
2. Screen lockers
3. Encrypting ransomware

Before we go further, I should note that this book is primarily interested in 'encrypting ransomware'.

Scareware is typically little more than malicious advertising. The user might see a pop-up advising that malware has been detected and instructing them to visit a website or make a payment to have the malware removed. This can be served on a website (in which case it's likely you're not infected at all and it's just an obnoxious pop-up) or from a malware infection on your device. In most cases, this is all that the malware will do, so it could be little more than an annoyance – but if you have this sort of infection, it's likely there's more malware that you just don't know about.

Screen lockers, meanwhile, look an awful lot like encrypting ransomware. They freeze up your device and often present a message stating that the user must either pay a ransom or that they are under investigation by the FBI or some other authority. This is obviously a more significant threat than scareware, but in most cases your data is otherwise safe – the malware is simply preventing you from accessing the device and trying to scare you into making the payment.

Encrypting ransomware is a much greater issue, because it's not possible to recover your files without either a great deal of luck (if the ransomware is old enough that there is a known solution) or by paying the ransom – and there's no guarantee that the criminals will hold up their side of the deal.

There's a growing subset of this in which the criminals will also take copies of your data. This usually acts as an additional threat in the ransom: if you don't pay up, not only will you lose your data, we'll release it onto the Internet; if you do pay, we'll unlock your files and dispose of the copies. Again, however, you have no idea if the criminals will actually delete their copies. If you know much about criminals, you'll join me in being sceptical.

There is a strong argument to never pay the ransom. Regardless of any ethical considerations, there's a simple practicality: you have no way of knowing if the criminals will actually unlock your information, nor any guarantee that they then won't mark you as an easy target (see the end of this chapter for more on the ransomware 'business model'). By far the best solution, as you'll see, is to be prepared so that you can prevent as many infections as possible, contain any ransomware that does make it through, and recover your systems and information quickly and with minimal interruption.

How it works

The core functionality of ransomware is two-fold: to encrypt data and deliver the ransom message.[10] Depending on the complexity of the malware and its mechanism for gaining access, the encryption can be relatively basic or maddeningly complex, and it might affect only a single device or a whole network.

Encryption is a relatively common feature in modern computing, and most users encounter it without ever realising. Most of the time, encryption is a valuable tool: it protects data from being accessed or interfered with by unauthorised parties. Not all information needs to be encrypted, but most people would conclude that it's valuable and often a 'better safe than sorry' solution, even for relatively mundane information.

An ideal system encrypts data to be sent (or stored) and the receiving device decrypts the information, all without the users necessarily being aware of the fact. This relies on cryptographic keys, which instruct how the information is encrypted and decrypted. In some cases, the same key can be used to encrypt and decrypt, while others use one key for each purpose. In any case, a key is necessary to access the information.

---

[10] There is a further subset of this: wipers. These overwrite the data in question, which renders it completely unrecoverable. The functional difference boils down to whether you intend to pay the ransom and trust the criminals to restore your data. If the malware is a wiper and the criminals didn't take copies, any ransom is simply throwing money away. It's likely you won't know the difference until it's too late.

Ransomware's main problem is that the victim has no access to a key to decrypt the data. Depending on the strength of the encryption algorithm, this can mean that the data is essentially unrecoverable. While some older algorithms can be broken given enough time and are still occasionally used in business either because the infrastructure is outdated or there's a perceived benefit to having simpler encryption, there's no need for a ransomware attacker to use this – they don't want the data to be readily accessible. For all practical purposes: if you're hit by ransomware, your data is lost unless you get the key or can recover it from somewhere that hasn't been affected.

Once the data has been locked up, the ransomware needs to notify the user, which is generally in the form of a pop-up or other notice on your screen, which will (possibly gloatingly) explain that your files have been encrypted and how to pay to get them unlocked. Depending on the scale of infection and the purpose of the ransomware, it might leave some systems accessible, even if in a limited capacity. For instance, the ransomware might lock down the device in its entirety, leaving only a message explaining how to pay the fee; alternatively, it might leave the user able to access the Internet so that the ransom can be paid.

One of these messages is usually the first sign anyone gets that they've been infected. So, how did it get there?

Mode of access

Ransomware, just like any other malware, needs a way in. This is generally called 'infiltration', and there are two primary methods:

 1. Social engineering, such as phishing.

2. Technical vulnerabilities in the network perimeter.

Social engineering often relies on human error – clicking links in phishing emails, allowing an unknown application to execute, and so on. The aim for any malware developer is to reduce the number of times they need humans to make errors: if they rely on someone to click a link, then accept a download and actively execute the file, that's three points at which the target might wake up and realise they're making a mistake. If the target only needs to click a link, the criminal's doing much better. If they can find a way to get the malware directly onto the target's device without them realising, they have an absolute winner.

Among the most popular methods is phishing, which has long been proven an effective way of slipping past technical defences. In this scenario, the target opens an attached document (probably Word or a similar format, with the malware payload delivered by a macro) or clicks a link, and they're promptly infected.

Ransomware can be delivered a number of other ways – via a wider infection once it gets into a network, for instance, within an infected USB device, packaged with a more 'benign' download (such as bundled with an app or other software from a disreputable source), and so on. As you can see, in many cases, the initial infection is due to human error.

It is also possible to deliver ransomware without human intervention, however, by directly attacking the network via vulnerabilities in the perimeter. Criminals are constantly probing network boundaries looking for the telltale signs of a vulnerability that they can exploit to gain access.

Undirected attacks against networks look for common flaws that can be automatically exploited. The WannaCry attack was able to propagate, for instance, by taking advantage of an NSA backdoor that had been recently revealed. There was a patch available, but too few machines had been updated.

More complicated attacks – clearly targeted – will combine data from a number of sources in order to gain access. For instance, responses from a login portal (such as making it clear if a specific username is correct even if the password is wrong) could be combined with information gleaned from LinkedIn (a list of employees) and previous data breaches (connecting a user with passwords they've previously used), which could give criminals a set of likely user credentials to attempt.

Other criminal groups sell known backdoors or details of previously compromised networks that ransomware attackers can target without needing to do too much of their own work to gain access. This access will generally have been gained the same way as those methods described above.

All of this is a way in, but ransomware also needs to establish itself without being detected. As malware is often single-purpose, ransomware can have some difficulty gaining a foothold anywhere, which is why it is often paired with one or two other malicious programs. As mentioned in the introduction, Emotet is a popular Trojan that does most of the legwork for ransomware, but there are others that do much the same.

The infection

The actual speed of movement once the malware is within the network can vary considerably, depending on the attacker's ambitions, the attack method and any internal technical security measures.

An attacker who isn't concerned about the specific data they're locking up will simply attempt to lock up as much as possible as quickly as possible until it's stopped. Indiscriminate, untargeted ransomware attacks are more likely to follow this approach. In this case, the ransomware will simply start grabbing files and encrypting them, often starting with files that aren't system-critical in order to minimise the risk of being spotted. It should be noted that this is a pretty primitive method, and most modern ransomware will take a more intelligent approach to maximise its impact.

A more complex ransomware attack will put measures in place to cripple defensive measures or hide from them, which can slow the rate of encryption – at least initially. This might include disabling antivirus and intrusion detection systems, system logs, and so on. The aim in these cases will be to maximise the amount that can be encrypted.

More advanced attacks will also attempt to spread through a network, not encrypting anything until as many systems as possible are carrying the infection. As with the initial entry into the network, this will be facilitated by other malware designed for the purpose. The ransomware itself will only be triggered once the malware has spread as far as possible. This activation might even be timed to minimise the chance that it's spotted in action, such as late on a Friday after the IT team has headed home for the weekend.

The most advanced attacks are supervised by the criminals. In these cases, the malware will snoop through as much as possible before triggering the ransomware, making sure that the most valuable data is affected, and applying more sophisticated methods to spread as far as possible while evading detection. Not only will your data and systems be locked up, but the criminals will almost certainly have copies of key data, access credentials, and so on.

Fallout

The most obvious, immediate impact of the encryption is that you will no longer be able to access or use the affected data and systems. This has an immediate, crippling effect. An infected computer, whether or not it has been completely locked down, cannot reasonably be used for anything. The further the infection has spread, the more profound the impact.

An encrypted sales database will leave you unable to fulfil orders or enter new ones. Locked down payroll systems will not only prevent you from paying your employees, but could also result in theft. Other effects in finance systems could prevent you from paying suppliers or processing invoices. Locked up industrial control systems can prevent industrial machinery from operating or remove safety protocols, making the work environment unsafe. Healthcare devices that rely on access to centralised information can cease working or revert to default settings, with potentially fatal results.

All of this leads to outcomes such as a loss of productivity, lost business, damage to the organisation's reputation, harm to clients, fines under data protection law, remediation and investigation costs, higher insurance premiums, the cost of remediation work to clean and rebuild the network, and so on. This is in addition to any ransom you choose to pay (which you shouldn't).

It's no surprise that most organisations are ill-equipped to deal with the resulting chaos, let alone continue functioning at any kind of 'normal' level.

How criminals use ransomware

One of the appeals of ransomware (and cyber crime generally) is that it is a very affordable type of attack. Like most malware, it is available cheaply online and can be deployed without much technical know-how. An attacker can easily purchase ransomware-as-a-service and access to previously compromised networks and start making money in fairly short order. Deloitte found that an average ransomware campaign would cost around US$1,000, which includes everything the criminal needs to get ransomware onto victims' machines.[11]

Unlike other kinds of malware attack, ransomware also has a clear business model baked in.

---

[11] Deloitte, 'Black-market ecosystem – Estimating the cost of "Pwnership"', December 2018, *www2.deloitte.com/content/dam/Deloitte/us/Documents/risk/us-risk-black-market-ecosystem.pdf*.

Ransomware is a very direct sort of criminal activity: unlike other cyber crimes that rely on getting access to information that the criminals hope will be valuable enough to convert into money, ransomware simply causes enough of a problem that a certain proportion of victims will choose to pay the ransom.

In order to maximise the chances of being paid, criminals can opt for a couple of strategies:

1. Hit as many targets as possible.
2. Pick a smaller number of targets but more carefully.

The first option is the common scattergun approach, much like ordinary phishing attacks, generally using purchased lists of email addresses harvested from data breaches and cyber attacks. People whose addresses have previously been put on these lists are thus more likely to be targeted again and again.

The second option, meanwhile, usually involves a bit of investigation to identify targets that are likely to pay the ransom and may have vulnerabilities that can be easily exploited. This approach might be combined with a wider operation to steal data or perform other types of attack to maximise the value for the criminals. As this is a targeted operation, the attackers will usually try spear phishing and similar techniques to get the ransomware inside. Because the attacker has already chosen to put extra effort into this type of attack, they're also more likely to make any such emails more convincing.

It is worth noting that criminals have tended to move towards the second option over time because the rewards are considerably greater. The WannaCry outbreak in 2017, for instance, was an example of an untargeted attack that simply got lucky, but it only netted around £100,000 for the attackers (while costing the rest of the world hundreds of millions).[12] Meanwhile, the small town of Riviera Beach in Florida paid US$600,000 to recover its information and systems.[13]

Eerily, many criminals treat their ransomware operations like something approaching a professional enterprise. They may have 'mission statements' claiming that they do not deliberately target certain types of organisation, such as hospitals or schools. They may offer 'live customer service' complete with polite representatives and a willingness to negotiate payment plans. They calculate their price points carefully, much like any other business. For instance, someone might think that the information on their laptop is worth £300, but aren't willing to pay more than that; a business, meanwhile, is going to value that considerably differently.

It is worth noting that criminal organisations that treat ransomware like a serious business are presumably more likely to honour their side of a ransom deal. It wouldn't be in their interests, after all, for their 'customers' to complain that cryptographic keys weren't supplied. In order to function as a 'business', they – like any other business –

---

[12] Samuel Gibbs, *The Guardian*, "WannaCry: hackers withdraw £108,000 of bitcoin ransom", August 2017, *www.theguardian.com/technology/2017/aug/03/wannacry-hackers-withdraw-108000-pounds-bitcoin-ransom*.

[13] BBC, "Florida town pays $600,000 virus ransom", June 2019, *www.bbc.co.uk/news/technology-48704612*.

should be bound by a duty to provide the services paid for. Trusting attackers to take this approach is not a decision to be made lightly.

CHAPTER 2: BASIC MEASURES

While this book sets out a programme that any organisation can implement to help protect itself from ransomware, there's a lot of crossover with the sort of measures you might put in place to protect yourself from cyber threats more generally. These should be in place in any organisation and form a strong foundation that can later be developed into something more comprehensive and mature.

The following measures will be built on in later chapters, so any organisation interested in an incremental approach can work up from first principles in an organic way.

Cyber hygiene

Cyber hygiene is about securing the organisation with simple measures. These are basic things that a lot of organisations already have in place to one degree or another but might not be treated consistently or given much attention.

As the majority of Internet-based threats target common vulnerabilities, simply plugging these gaps can considerably improve security. It's not a simple fire-and-forget solution, however, because the Internet is a constantly evolving battleground. Cyber hygiene is something that needs to be monitored and reviewed, much like any other business activity: if something isn't working, or costs too much, you should be checking to see how it can be improved.

In the UK, the Cyber Essentials scheme represents a good approach to basic cyber hygiene. It posits five key controls:

1. **Firewalls**

   This addresses the general need to protect the network boundary from Internet-based threats. Ensuring firewalls are in place, correctly configured and managed appropriately can prevent a host of common threats from gaining access to the network. Access to the firewall's administrative interface should also be restricted so that it can only be configured by approved admins.

   The rules that firewalls use to permit inbound and outbound traffic should be reviewed and approved by an appropriately authorised person. These rules should also be justified in business terms. For instance, if a rule permits a relatively broad swathe of traffic, there should be a compelling business reason to do so.

2. **Secure configuration**

   All software and hardware come with a default configuration. This is usually assigned on the basis that it improves accessibility or usability for the software, but it isn't always the most secure configuration.

   In many cases, software and devices will also offer services that aren't necessary for the business. As each instance of this presents a potential vulnerability, they should be disabled by default. Equally, default passwords should be changed and default access rights should be reviewed – after all, it's unlikely that every user has the same access needs.

There's obviously a balance to strike when configuring software and hardware, as it's easy to lock down an application or system to such a degree that it's no longer useful.

3. User access control

The organisation should have robust methods for verifying the identity of anyone trying to access information or systems, and each user should have strictly defined access rights.

Passwords are an obvious part of this, but some systems may warrant multifactor authentication as an additional security measure.

Privileges should also be carefully considered on the basis of need. If you only need to access a system very irregularly, then you don't need to have perpetual access to it; this sort of access should be granted for a defined time period when it's necessary, then revoked.

All of this should be built into an effective user account management process. This process should ensure that when user accounts are first created they have minimal access rights and effective user verification, and that access privileges and rights are reviewed when the user changes roles, and again when the user leaves or is terminated. Admin accounts should obviously be strictly controlled and prohibited from being used for 'ordinary' duties.

4. Malware protection

Malware protection can be achieved in a few ways: anti-malware software, application whitelisting and application sandboxing.

Antivirus and other such applications and systems should be kept up to date and protected from interference. They should be configured to scan files upon access, and web pages when browsing.

Application whitelisting should be restricted by code signing to prevent unpermitted applications from executing. Lists of approved applications should be reviewed periodically to ensure only secure applications with a business purpose can be executed.

Application sandboxing should ensure that 'unknown' applications can only be run in a sandbox until they are verified as secure and approved for use on the wider network. This should apply to all applications and hardware devices.

5. Patch management

All software, applications and systems should be maintained to eliminate vulnerabilities as quickly as possible. This includes ensuring that software, applications and systems are licensed and supported. When no longer needed or support is no longer available, they should be removed.

The organisation should also assess the criticality of new patches and schedule them accordingly. For instance, where a patch resolves a critical or high-risk vulnerability, it should be prioritised.

The Cyber Essentials scheme is backed by a certification programme that can provide some assurance that these measures have been appropriately implemented. For organisations new to cyber security, it therefore provides a good basis for development and building expertise.

Beyond the technical controls necessary to protect the organisation, cyber hygiene also involves recognising the scope of the extended network. Your organisation's network probably isn't just the computers and other hardware physically in the office: it may include staff mobile devices, company laptops and tablets, managed IT services, Cloud services, and so on. All of these need to be considered because they're effectively part of your network and a ransomware attack on any one of them could leak into your office network. While you might not be able to implement controls on services provided by third parties, contracts should establish that certain security measures are in place.

Staff personal devices can be a tricky point to manage, as few people are happy with their employers dictating how their devices are used or insisting on certain applications being installed. However, as these devices are among the most likely to be stolen or misplaced, it's important to ensure that the organisation has measures in place to protect itself. This might include requiring users to have PIN access to the device (for a start), requiring anti-malware, only granting access to company networks and applications via VPN, and so on.

Staff awareness

As established in the previous chapter, a lot of the ways ransomware (and other malware) finds its way into your networks is through human error. It is crucial to ensure that all staff have a basic understanding of cyber security, how to identify tricks like phishing emails, and how they contribute to the organisation's security.

Staff awareness is a huge field, and there are dozens of approaches you can take. These include having dedicated training sessions, using online training courses, e-learning, posters and other reminders around the office, and so on.

Whatever approach you take, you should ensure that users are reminded about cyber security on a regular basis. More intensive training can be relatively periodic – when an employee starts, then every six months or annually, for instance – but it's easy to forget about things. Reminders don't need to be full of effort, either: a quick all-staff email to remind them about the threat and perhaps mentioning an organisation in the same sector that succumbed to cyber crime might be perfectly adequate.

Effective staff awareness training needs to target multiple mediums and be repeated regularly. Research on phishing and security staff awareness training has found that people tend to forget what they've learnt from such courses after three to six months.[14] Of course, simply repeating the same training time and time again isn't likely to have the best impact. Users will get bored, for one, and changes in criminal activity and technologies can leave the content out of date. The material should be regularly refreshed for accuracy, and feedback from users can help improve retention and alleviate boredom.

Part of staff awareness is also making sure people know about the organisation's efforts to combat cyber crime. This extends to providing clear policies and procedures that all staff know how to find when they need them.

Surviving ransomware

If your organisation is struck by ransomware (or any other cyber attack, for that matter), you'll want to know that you have prepared and can recover quickly. Knowing what you can do ahead of time is critical.

Part of this is understanding the steps you'll need to take to deal with the infection. In broad terms, you'll probably go through the following stages:

1. Lockdown and contain the infection
2. Report to authorities
3. Clean your devices
4. Restore software and data
5. See what lessons you can learn to prevent it recurring

---

[14] Usenix, "An investigation of phishing awareness and education over time: When and how to best remind users", August 2020, *www.usenix.org/system/files/soups2020-reinheimer_0.pdf*.

Locking down and isolating an infection as quickly as possible will help minimise the amount of work you need to do later. In many cases, this will be as simple as disconnecting the device from all network connections (both wired and Wi-Fi) and identifying any other devices that might need to be isolated. If you've identified the specific ransomware in action, this might give you some idea of how far the infection could have spread. It's crucial to make sure you identify the full scope of the infection and isolate it, otherwise any restored devices will simply be reinfected over and over again.

Reporting ransomware incidents to authorities may be a legal requirement, depending on the type of data affected and your sector. A lot of authorities offer support and guidance in responding to cyber crime – for instance, the UK's National Cyber Security Centre (NCSC) has extensive information about ransomware, how to prevent it, and links to content that may be able to significantly mitigate the impact. Law enforcement agencies are also likely to be interested in the attack and any evidence you might recover that could identify perpetrators. In some cases, they may provide forensic support to gather specific evidence.

'Cleaning' infected devices usually means completely wiping the device and any associated storage. While this can be painful, it's absolutely critical because any remaining infection could simply re-emerge and propagate if the device is reconnected to the network. Furthermore, the impact will be lessened by the next stage, assuming you've made the appropriate preparations.

Clean devices can have software and data restored, as long as they are from a secure source. For this reason, the organisation should have a clear backup policy and processes that ensure any data lost to ransomware is minimised. The frequency of backups should be determined by assessing the maximum acceptable loss. For instance, if the organisation would be dangerously impacted by losing a day's data, the backups should be taken more frequently. This should also take into account restoration – if it takes too long to restore the data from backup (if the backups are stored on physical media at a separate site, for instance) they may need to be taken more frequently to account for this delay.

The backup and restoration process should be tested regularly, especially if there are frequent changes to systems or processes.

Finally, once the ransomware has been purged and business operations are getting back to normal, you should review the process. What did you do wrong? What worked well? Where are the pain points? How can the organisation reduce the chances of the incident recurring? The results of this review should be fed back into your defence and response measures.

A lot of these activities can be prepared for relatively easily with a little thought. You can develop processes for identifying the extent of an infection and isolating affected devices. You can implement automated and manual backups and the processes for restoring them. You can think ahead about the sorts of things that could be useful to make note of for the review. One of the keys to making sure they all work on the day, however, is practice. Rehearsing for a ransomware attack will help you identify and resolve problems and oversights before an actual attack.

CHAPTER 3: AN ANTI-RANSOMWARE PROGRAMME

The first step beyond having simple measures in place should be to make your efforts systematic. The programme we describe in this book is fundamentally a set of controls to reduce the risks posed by ransomware, but it should be implemented as part of a larger structure to ensure that these controls are subject to review and revision on an ongoing basis.

The first part of this will be to ensure there is a single person responsible for the project who has direct access to the top of the organisation. As the programme needs to cover potentially a great deal of the organisation and will require some investment, this will be critical.

Your approach to implementing the programme will, in many ways, define how it operates as a function of the organisation. At an early stage in the implementation, you should build in processes to monitor and review the controls. Having this in place from the start means that you can be refining your first controls while you're implementing later ones, building good experience and improving your capabilities at the same time.

In order to achieve this, you'll need to establish suitable metrics for assessing each control. As we discuss the controls in the following chapters, you should get a feel for how they can be implemented in your organisation and the metrics you might use to assess them. It's worth remembering that you're not just measuring how effective a control is, but also how well it has been implemented. A control might be effective, but if it's onerous or expensive, there's still room for improvement.

By way of example, a control to provide staff awareness training might be measured by assessing the proportion of staff who complete the training, scores in any tests, and the number of phishing emails that users fall for. Metrics to assess the implementation might measure how long it takes users to complete the training, the cost of any trainers or learning materials, and feedback from staff.

These processes should also be applied to the programme as a whole, with the person responsible for the programme reviewing the overall progress and effectiveness on a regular basis so that they can report the results up the chain to the top of the organisation.

The control framework

Table 1 on the following pages shows the structure of the control framework. It indicates which controls should be considered for organisations of varying sizes, addressing small and medium-sized enterprises (SMEs) as distinct from all other organisations, and then providing additional controls for organisations at high risk of ransomware attacks or likely to suffer considerable impact.

You might consider yourself at high risk if you've previously been affected by ransomware and paid the ransom, or if you are subject to ongoing ransomware attacks, perhaps due to your industry or media attention. High impact is likely to be determined by factors such as legal or regulatory requirements to protect information, or the impact on customers or the public. For instance, critical infrastructure organisations might be relatively secure financially, but a ransomware attack could have an overwhelming impact on the public.

Controls are divided into categories according to the type of activity and the overall objective. Each control provides a simple run-down of its objective, which we'll explore in more detail in the following chapters.

Table 1: The Control Framework

Ref.	Control	Description	SME (low impact)	All other	High risk / impact
1.0	User training	**Educate users to prevent malicious software being run by user interaction.**			
1.1	Anti-phishing training	Train users to detect phishing and suspicious emails.	✓	✓	✓
1.2	Phishing testing	Test users' responses to phishing emails.		✓	✓
1.3	Incident reporting training	Train users on how to report an incident.	✓	✓	✓
1.4	Incident response training	Train users on how to respond to incidents.			✓

Ref.	Control	Description	SME (low impact)	All other	High risk / impact
2.0	**Email**	**Configure email to reduce malicious messages arriving at users' mailboxes.**			
2.1	Email authent-ication and verification	Implement SPF, DKIM, DMARC.		✓	✓
2.2	Email filtering within mail servers	Implement email filtering such as Office 365 anti-phishing protection, ATP filtering, junk filtering, safe links, safe attachments.	✓	✓	✓
3.0	**Web filtering**	**Prevent access to known malicious sites and prevent drive-by ransomware downloads.**			
3.1	Block malicious sites	Firewall or browser-based tools.	✓	✓	✓

Ref.	Control	Description	SME (low impact)	All other	High risk / impact
3.2	Prevent malicious downloads	Firewall or browser-based tools.		✓	✓
3.3	Maintain blacklists	Keep browser and antivirus blacklists up to date.			✓
4.0	**Anti-malware**	**Install anti-malware to detect ransomware on hosts.**			
4.1	Install anti-malware on hosts	Install a reputable anti-malware solution on all hosts if it is available (workstations, servers, etc.).	✓	✓	✓
4.2	Use centralised monitoring and reporting	Enable remote monitoring of host devices from a central location.		✓	✓
4.3	Keep anti-malware engine and signatures up to date	Enable auto-update of signatures and anti-malware software.	✓	✓	✓
4.4	Enable anti-	Windows® Defender anti-		✓	✓

Ref.	Control	Description	SME (low impact)	All other	High risk / impact
	ransom-ware technology	ransomware not enabled by default.			
4.5	Implement filename spoofing prevention	Show file extensions, prevent techniques like RTLO or RLO.			✓
5.0	**Back-up**	**Protect data in case of a ransomware attack so it can be restored.**			
5.1	Determine location and RTO of critical data	Determine how much data can be lost before it affects the organisation's operations and inform backup strategy.			✓
5.2	Implement appropriate backup strategy	Implement a back-up strategy to ensure important or critical information can be restored in case of attack.	✓	✓	✓
5.3	Keep at least one copy of backup	Ensure at least one copy of data cannot be accessible from	✓	✓	✓

Ref.	Control	Description	SME (low impact)	All other	High risk / impact
	offline or segregated from hosts	hosts that could be infected with ransomware.			
5.4	Implement a back-up testing strategy	Test backups to ensure they are good and can be restored.	✓	✓	✓
6.0	**Network**	**Configure the network to reduce the lateral travel of ransomware.**			
6.1	Segregate network based on common security require-ments	Use VLANs or separate network to reduce the scope of an attack.		✓	✓
6.2	Prevent workstation -to-workstation traffic	Reduce the ability of ransomware to spread between workstations.			✓
6.3	Implement secure protocols	Remove insecure protocols used by ransomware to move laterally, i.e. use SMBv2 instead of SMBv1.		✓	✓

Ref.	Control	Description	SME (low impact)	All other	High risk / impact
6.4	Implement only necessary traffic between network segments	Reduce the ability of ransomware to spread across network segments.			✓
6.5	Implement MAC/IP anti-spoofing protection	Implement measures to prevent ARP or DNS spoofing attacks to prevent redirection to malicious sites.			✓
6.6	Review permissions on shared drives	Reduce the ability of ransomware to attack shared network drives.	✓	✓	✓
7.0	**Perimeter security**	**Prevent ransomware communicating with command and control servers (C&C).**			
7.1	Block access to C&C networks	Implement blacklists based on known C&C sites.			✓
7.2	Subscribe to	Keep blacklists of C&C servers up to date by			✓

Ref.	Control	Description	SME (low impact)	All other	High risk / impact
	intelligence feeds	subscribing to a security feed.			
8.0	**Access control**	**Prevent malware from using privileged accounts to move laterally.**			
8.1	Use separate admin accounts	Reduce the chance of a privileged account being used by ransomware.	✓	✓	✓
8.2	Use admin accounts only for activities that require that level of access	Do not use admin accounts for accessing the Internet to reduce the chance of infection.	✓	✓	✓
8.3	Implement MFA for admin accounts	Prevent the brute-forcing of credentials.			✓
8.4	Limit user permissions to the minimal	Implement business-need-to-know and least privilege to prevent users accessing the whole network.		✓	✓

Ref.	Control	Description	SME (low impact)	All other	High risk / impact
9.0	**Host hardening**	**Harden hosts to reduce the likelihood of infection by ransomware.**			
9.1	Disable autorun	Disable autorun on removable media, network, web downloads to prevent ransomware from spreading.	✓	✓	✓
9.2	Prevent applications running from user-writable areas of the hard drive	Prevent ransomware installing via drive-by download and user interaction.			✓
9.3	Keep OS and applications up to date	Prevent ransomware from using known software vulnerabilities.	✓	✓	✓
9.4	Remove unnece-ssary applications	Prevent ransomware from using known software vulnerabilities.	✓	✓	✓
9.5	Use OS and applications that are	Prevent ransomware from using	✓	✓	✓

Ref.	Control	Description	SME (low impact)	All other	High risk / impact
	supported by vendors	known software vulnerabilities.			
9.6	Disable macros in applications	Prevent malicious macros from running.	✓	✓	✓
9.7	Define software restriction policies	Prevent malicious applications from running.		✓	✓
9.8	Disable Windows Script Host and PowerShell	Prevent ransomware from leveraging built-in tools to spread.			✓
9.9	Whitelist applications	Prevent malicious applications from running by deploying tools such as AppLocker application control policies.		✓	✓
9.10	Run unknown applications in a sandbox	Web applications to be run in sandbox unless approved by the organisation.		✓	✓
10.0	Detection	Detect the operation of ransomware so			

Ref.	Control	Description	SME (low impact)	All other	High risk / impact
		host can be cleaned before encryption is complete.			
10.1	Monitor for known malicious file extensions	Look for known extensions used by ransomware.			✓
10.2	Monitor for increase in file renames	Monitor file renames as ransomware encrypts files and adds its extension.			✓
10.3	Implement sacrificial network shares/data files	Create honeypots for detecting ransomware activities.			✓
10.4	Implement IPS/IDS with exploit kit detection	Implement IPS/IDS that can detect the presence of ransomware.			✓
10.5	Monitor processes that read or write too many files too quickly	Monitor for processes that are accessing large numbers of files in a short time.			✓
10.6	Deploy file integrity/	Look for changes to files			✓

Ref.	Control	Description	SME (low impact)	All other	High risk / impact
	file change monitoring software	caused by ransomware.			

Risk management

While we provide a comprehensive set of controls for organisations of different sizes and with different needs, it would be foolish to assume that this is all you need to do – or that you need to implement all of the prescribed controls precisely as described. For any given organisation, some controls will be more or less valid: if you have a very small, highly tech-literate staff, for instance, you may not need to put as much effort into anti-phishing training (control 1.1) because you can rely on your staff to be aware of the threat. Equally, however, you might need to make sure that every member of staff can support the incident response activities (control 1.4) because of the size of your organisation.

Determining the right mix of controls and the necessary maturity of those controls should be based on some form of risk management. Larger organisations should have a risk management function and can easily assess their own needs, while SMEs may benefit from a streamlined approach that assumes each of the SME controls is necessary and then examines the necessity of the more advanced controls on a case-by-case basis.

Remember that the anti-ransomware programme should be reported on to the top of the organisation, so decisions to expand the programme or implement expensive measures will doubtless need to be justified.

Controls

Each control category and control has a simple description, which should inform how you determine the success of the implementation. That is, if a control is not meeting the description, then it needs to be reviewed and corrected. Equally, if the set of controls within a category do not meet the category's description, then one or more controls may need to be adjusted. You could develop these descriptions into objectives for the controls and categories to inform how you monitor and review their effectiveness.

Each control should have an 'owner', which is the person responsible for making sure the control is implemented and adhered to. In many cases, that will be the person responsible for the programme, while in other cases it's useful to assign ownership to someone closer to the processes that the control affects. For instance, the 'disable autorun' controls (9.1) would be usefully assigned to someone within the IT function who is responsible for configuring hardware and software.

The controls are individually described in the following chapters, along with basic guidance on implementation and improving maturity.

Maturity

Each control should mature over time, becoming more and more part of your organisation's ordinary operations and increasing in effectiveness. The pace at which you approach maturity will depend on a number of key factors:

1. How seriously the organisation takes the threat.
2. The dedication of top leadership.
3. The diligence of everyone involved.

While you should be aiming to develop more mature capabilities, you will not be able to achieve that goal across the board without these factors in place. It's clear that having an engaged workforce (at all levels) will provide strong support for the programme, so it's worth making sure you have processes in place to develop this support. As a minimum, the programme should be introduced to new starters and reiterated to staff on a regular basis. A lot of this will overlap with the user training controls set out in the following chapters.

Assessing and improving maturity can be approached in a number of ways, and there are common models with plenty of supporting materials available. The NIST Cybersecurity Framework, for instance, describes 'implementation tiers' with simple definitions so the organisation can quickly identify its level of maturity and how far it still has to go.[15] Definitions used in other frameworks may also be suitable, especially if the organisation already has some experience with them.

It's also worth remembering that, in some cases, greater maturity comes with increased costs or effort, and the return on investment may not support that. The current maturity and desired or necessary maturity should be

---

[15] National Institute of Standards and Technology, *Framework for Improving Critical Infrastructure Cybersecurity*, April 2018, *https://nvlpubs.nist.gov/nistpubs/CSWP/NIST.CSWP.04162018.pdf*.

assessed as part of the overall review of the anti-ransomware programme.

CHAPTER 4: BASIC CONTROLS

These controls have been selected as suitable for SMEs and can be implemented in every organisation at relatively little cost or effort. While they may not all be necessary for every SME, they can protect an organisation as it grows and should be considered best practice.

Larger organisations likely already have many of these in place, but should verify that they are and that they are appropriate to the organisation's size and operations.

1.1 Anti-phishing training

Because phishing is one of the most common ways for malware to enter the organisation, all staff should be trained to identify it and understand what they should do when they receive phishing emails. Training should also consider other ways in which people might be subject to phishing attacks, such as through messaging apps, text messages, and so on.

Training can be provided in whatever format is most suitable: formal classroom-based training, e-learning, remote learning, and so on. This should be backed up by reinforcement, such as through posters in the workplace, company briefings, regular testing, reminder emails or newsletters.

As with any awareness training, you should ensure all staff get the training when they first start and at regular intervals throughout their employment. It should be extended to anyone else who has access to the organisation's systems, such as contractors, temporary workers, suppliers, partners, and so on.

A more mature approach to such training will assess the effectiveness through activities such as simulated phishing attacks (see control 1.2 in the next chapter) so that the training becomes more relevant to the organisation's activities and, hopefully, more effective as a result.

1.3 Incident reporting training

All users should know how to report an incident, which can extend beyond just ransomware attacks. As ransomware attacks can render the user's devices useless or hazardous, training should explain how a user can report an incident when their workstation or laptop (or other primary device) may have been affected.

The incident reporting process should ensure that the right information is available to response teams as quickly as possible. You might implement this as part of a service desk function, backed up by phone and email, with forms to complete to gather key information.

In most cases, staff don't need to know off the top of their head exactly how to report an incident or the specific information they need to provide. However, they do need to know how to find out how to report an incident. A good incident reporting system is documented in a common-sense location. If users know where all of the organisation's process documentation is kept, that's where information about the incident reporting process should also be kept. If your IT ticketing system is where users often report other types of incident, you should incorporate an incident reporting function there.

2.2 Email filtering within mail servers

Mail server-level filtering applies a set of policies to inbound and outbound emails, blocking those that exhibit certain traits so they never reach the intended recipient. This can effectively prevent a large volume of phishing emails, as well as potentially malicious attachments, from reaching the target. User-level filtering (such as setting up filters in Outlook to move suspicious emails to the trash) is less effective because the user still has access to the potentially malicious email and could – accidentally or otherwise – execute malware or follow malicious links.

Rules for outbound traffic can also be configured to reduce the chances of sensitive information being disseminated. In the event that your networks are infected, it can also prevent further email-borne infections originating from your organisation.

Very nearly every email provider now includes filtering tools, such as Office 365's anti-phishing protection, which can be customised to meet your requirements. Filtering should also address junk mail and spam, and can be set up to address risks in content that is let through the filter, such as by applying safe link and attachment policies that scan links and attachments to ensure they are safe.

The emails blocked and rules you apply should be reviewed with some regularity to ensure they are strict enough without preventing legitimate email from passing through.

Rules will usually be based on analysis of the header data, but smarter criminals will disguise this information to bypass common rules. These rules can be bolstered by referring to blacklists of known spammers and cyber criminals, which can be procured from anti-spam providers such as Spamhaus.[16]

3.1 Block malicious sites

Users should be prevented from visiting malicious sites by either a firewall or browser-based tools (ideally both). Most browsers will now warn users if a website they are about to visit is suspicious, and company-wide user policies can prevent users from accessing any website that triggers these warnings.

Firewalls can also apply blacklists to prevent access to malicious sites and inappropriate content, while a more restrictive approach might favour whitelists. Blacklisting on the basis of suspicious events and known malicious domains reduces risk while leaving the Internet largely open to users; whitelisting known safe domains can result in a more restrictive environment but protects business-critical websites and services. A combination of the two – using a list of known malicious domains and dynamically blocking domains that exhibit suspicious traits while protecting known safe domains – reduces browsing risks considerably while protecting business interests.

Both browser- and firewall-based methods should be used to minimise the risk of users encountering and engaging with malicious content.

---

[16] *www.spamhaus.org/*.

4.1 Install anti-malware on hosts

All hosts should have anti-malware software installed. The software should be configured to examine content online, as well as scanning unknown files and applications before permitting them to be opened or executed. It should also scan the host at least daily for any malware that may not have been identified under previous signatures.

Anti-malware solutions can also be layered to provide more comprehensive protection without putting too much load on the hosts. You might achieve this by running one application constantly with a second only running intermittent scans. Alternatively, different anti-malware and antivirus suites can be combined, with conflicting functions enabled or disabled to get the best of both worlds.

You should also be aware that different hosts have different needs for anti-malware, and that virtual and Cloud-based hosts also need this protection. One anti-malware solution isn't likely to be ideal in every situation.

4.3 Keep anti-malware engine and signatures up to date

As with any other software, anti-malware needs to be kept up to date. This applies to both the software engine and the signatures it uses to identify malware.

While some software updates may need to be reviewed for suitability and the presence of vulnerabilities before installing patches, anti-malware should be generally exempt from this policy. In order to spot and block the latest malware, the software needs to use the latest capabilities and signatures. Most anti-malware software is given daily signature updates, so it should be configured to accept and apply these updates as and when they are provided. Engine updates are less frequent but should be given similar priority.

This means that you should prevent users from interfering with these updates. To minimise impact on productivity, you might schedule engine updates for evenings or other low-usage times.

Because some malware will attempt to interfere with anti-malware updates, you should ensure that updates are occurring for every host. Any host that isn't updating signatures or engine patches should be investigated to determine whether malware is to blame.

5.2 Implement appropriate backup strategy

The backup strategy sets out the organisation's objectives for backup and restoration of data and systems. A simple approach might be to use file synchronisation to ensure that data is copied to a Cloud service (such as OneDrive) as and when changes occur, but this has limitations in that it won't necessarily protect backed-up data from accidental deletion or ransomware. A more mature approach will consider matters like the value of different data sets and more specific requirements for certain data (see control 5.1 in the next chapter).

For SMEs, the simplest approach is to simply determine a baseline approach to backup and recovery. This might be as simple as 'all data and systems are backed up daily', or take a more nuanced approach like 'data used in key functions is backed up daily; other data (such as email, etc.) is backed up weekly'.

Regardless of your approach, you will need to establish a system for taking backups that cannot be interfered with or blocked, and a process for restoring information from backup when necessary. There may also be legal considerations for some types of data. For instance, personal data is covered by a number of laws around the world, and depending on your backup methods may need to be treated differently from other types of data.

5.3 Keep at least one copy of backup offline or segregated from hosts

Because backups are necessary in the event of a ransomware attack, you should ensure that at least one copy of backup data is not accessible to the hosts that might be infected. Recovering from a remote backup can take more time and effort than backups stored on the local network, but it provides valuable redundancy.

The classic example would be to write backups to CD or external storage media, but that's a slow and inconvenient process, and creates issues with the physical storage space necessary for comprehensive backups.

A more practical solution would be using temporary network links to a backup server that is otherwise segregated from the hosts, or using an escrow system to transfer the backups to a Cloud repository. This means that the backups can be taken relatively easily while protecting them from infection.

5.4 Implement a backup testing strategy

Backups are only valuable if you can recover data from them. They have some value – albeit diminished – if it's difficult to recover data from them. You should regularly assess the effectiveness of backups by testing your recovery process. Depending on your method of storage, you may also need to assess how well the information is preserved (which can include making sure that your process recovers the right version of a backup).

If backups are encrypted, you also need to ensure that you are able to decrypt them. This part of your recovery process needs to be thought through to make sure the keys will be accessible under a range of possible conditions. The same is reasonably true of access rights to backups: if backups can only be accessed from specific devices or accounts – which could be compromised – you need to ensure there's a secure alternative.

6.6 Review permissions on shared drives

Shared drives provide valuable access to useful information across different business functions, but ransomware can easily exploit write permissions to both increase the volume of data that can be encrypted and spread across the network.

Users should only have write permissions for sections of shared drives that they need to be able to edit. Other sections should either be read-only or inaccessible. If a user has no need to edit any data on a shared drive, they should be limited to read-only access.

Access rights should be reviewed regularly, and anyone granted unusual rights should have an expiration date on those rights.

8.1 Use separate admin accounts

Administrators should have separate admin and user accounts. In conjunction with control 8.2 below, this reduces the likelihood that an administrative account is compromised, and thereby limits the ransomware's ability to spread through the network.

While the non-admin account may have elevated privileges in some areas, these should be controlled in the same way that any other user's privileges are limited. The administrator should only have cause to use their admin account for actual administrative purposes (see 8.2 below).

8.2 Use admin accounts only for activities that require that level of access

Admin accounts should only be used for admin purposes. If the current task does not need admin privileges, it should be carried out on the administrator's ordinary user account. It would be useful to have a matrix setting out what can only be done on an admin account so that it's clear when the administrator should be using it and when they should use their normal account.

The admin account should also be prevented from accessing the Internet or email. This minimises an admin account's contact with Internet-borne infections. Normal, day-to-day duties should be carried out on the user account.

9.1 Disable autorun

Autorun functions should be disabled for removable media, network devices and web downloads. For obvious reasons, this helps to limit the spread of malware.

If a system needs to retain autorun functionality for some reason, the justification should be documented and the conditions around it carefully constructed to reduce the opportunity for it to be exploited.

9.3 Keep OS and applications up to date

As previously discussed, nearly all software and operating systems will have a number of vulnerabilities, even if they've not been discovered yet. The vulnerabilities can be minor or severe, but the larger the volume of vulnerabilities the greater the risk to the organisation. As organisations use a wide array of software, the number of vulnerabilities that could be present grows rapidly.

While unknown vulnerabilities may present very little immediate risk, those that have been published can present a significant risk if they remain unpatched. When patches are released for software and operating systems, they should be assessed and prioritised. Patches for high-risk or critical vulnerabilities should be applied as soon as possible, while patches for less significant vulnerabilities may be reviewed to identify any issues they may create (such as changes to the way the software operates or creating new vulnerabilities related to the way the software is used) before being rolled out.

The organisation should also monitor feedback on patches to determine whether any new vulnerabilities have been introduced that may be treatable until a further patch is released.

Updates and patches should be managed centrally to ensure they are rolled out across the network consistently as and when necessary.

9.4 Remove unnecessary applications

As all software presents a potential security risk, the number of pieces of software present on the network has a direct correlation to the number of vulnerabilities present. It is sensible, then, to reduce the number of software items on any given device.

Software that the organisation does not use should be removed. This eliminates the need to apply patches as well as removing any vulnerabilities that may affect the software.

This software should be removed from the images used to configure new devices to reduce the risk that a copy slips through into the network.

9.5 Use OS and applications that are supported by vendors

Unsupported software will no longer receive patches for vulnerabilities, so should be removed and replaced by software that is still under support.

Most vendors will announce ahead of time when support will be removed, which is when the organisation should start looking for a replacement. In many cases, the vendor will have a newer version of the software available long before ending support.

Because changing an operating system across the organisation can be disruptive, it should be planned well to minimise disruption. Needless to say, any software that will need to run on new operating systems should be assessed to make sure that:

a) It will continue to function as intended; and
b) It is not subject to any unresolvable security vulnerabilities.

9.6 Disable macros in applications

Macros introduce executable code into files that normal users will generally consider benign, such as Word documents, PDFs and Excel spreadsheets. Disabling macros (and preventing users from enabling them) prevents these being used as vectors for delivering malware.

However, macros are often also used for valid purposes, so the organisation may need to permit some uses. For these exceptions, there are a few options:

1. Disable macros in all applications except those that you know require them.
2. Only enable macros for the staff who rely on them.
3. Disable macros unless they have been digitally signed by a trusted source.

A better solution is to reduce reliance on macros entirely, such as by replacing their functionality with either off-the-shelf plug-ins or by building applications.

Some anti-malware solutions can also scan macros to detect functions and behaviour that indicate they are likely to be malicious.

CHAPTER 5: ADDITIONAL CONTROLS FOR LARGER ORGANISATIONS

The following controls are suited to larger organisations as additional measures on top of those set out in the previous chapter. In many cases, they build on capabilities that should already exist or are developed in the earlier controls.

1.2 Phishing testing

Phishing training is all well and good, but if you decide that the measure of success is that the organisation has staved off ransomware attacks, you're running the very real risk that the training hasn't been effective. It's also worth remembering that a ransomware attack may only need one person to slip up.

A low-risk way of assessing the effectiveness of your training is to run phishing simulations. These fire off fake phishing emails at staff and record the responses. A basic test might use a fairly obvious phishing email template, complete with terrible spelling and grammar, obviously fake URLs, and so on, while a more difficult test might clone a company email template, use links designed to mimic authentic company domains and appear to follow standard company processes.

Measuring how many people fall for the email at each stage (such as clicking links and/or providing information) against the number of people who respond appropriately (reporting the email, etc.) will give you a good idea of how effective the training has been. It can also give you relatively good data about where your risks are likely to be and who might need additional training.

Simulated phishing attacks often require specialist skills, which you're unlikely to have on hand, but many penetration testing suppliers also offer simulated phishing services.[17]

2.1 Email authentication and verification

Email headers can be spoofed relatively trivially which can make it more difficult to automatically filter phishing emails. Implementing three key protocols on mail servers to authenticate and verify emails can help minimise this risk.

Sender policy framework (SPF) enables the mail server to determine whether a message originates from the domain it claims to, which can help reduce email spoofing. This does not protect against spoofing of the visible email address, however, so phishing emails that make it through may still fool users.

Domain keys identified mail (DKIM), meanwhile, checks digital signatures to confirm that an email was actually sent from the domain that it claims. It can also be used to verify that certain elements of the email (such as attachments) have not been modified since the signature was attached. This can help prevent phishing emails that claim to be from other known sources, as well as cruder attacks that attempt to mimic forwarded emails with malicious attachments.

---

[17] See IT Governance's service:
www.itgovernance.co.uk/shop/product/simulated-phishing-attack.

Finally, domain-based message authentication, reporting and conformance (DMARC) acts as an extension of SPF and DKIM by essentially publishing information about the mail server's email protocols so that other organisations can validate emails that come from the domain. This can protect the organisation from being used as a 'patsy' in other phishing attacks and encourages good practice more widely.

3.2 Prevent malicious downloads

Malicious downloads can be blocked by both firewalls and using browser-based tools. Firewall rules should be configured to prevent downloads of the obvious kind – executables – and from suspicious domains. It may be perfectly valid to prohibit downloads more generally, although this can limit access to some useful content, such as PDFs.

Browser plugins can prevent malicious downloads in a number of ways, including by preventing JavaScript from running, blocking advertisements, and so on, and by relying on the browser's built-in 'safe browsing' features. These should be enabled in the images used to configure devices, and you may wish to limit a user's ability to disable them.

4.2 Use centralised monitoring and reporting

Hosts should be monitored for suspicious activity from a centralised location, which will give a clear overview of any spread in ransomware or other malware. Obviously, manually monitoring log files is difficult, to say the least, so the organisation should implement a security information and event management (SIEM) solution to automate the process around the clock and provide alert notifications that can then be checked and followed up on.

The organisation should monitor a range of activity, across networks, systems and services. Inbound and outbound network traffic should be assessed for unusual activity, such as connections from unexpected IP ranges, large data transfers, and so on. User activity should also be monitored to identify unauthorised or accidental misuse of systems or data. The critical aspect of this is that it can tie a specific user account to suspicious activity, which can also help trace the source of infections. There is obviously some balance required, as there may be legal and regulatory constraints on monitoring users, not to mention potential impact on employee satisfaction and productivity.

4.4 Enable anti-ransomware technology

Windows and other operating systems come with some anti-ransomware capabilities bundled in, but they are not enabled by default. Microsoft® Defender has a range of ransomware protection, for instance, such as controlled folder access, which protects controlled folders by only allowing trusted applications to access them.

Controlled folder access events can be reviewed in Windows Event Viewer, which will highlight apps that have been prevented from accessing controlled folders and may provide evidence of a ransomware infection. These reports can be acted on as and when they arise, giving you a good opportunity to get ahead of a ransomware infection before it gets out of hand.

6.1 Segregate network based on common security requirements

Network segregation is an effective measure for limiting how far ransomware can spread, while protecting critical information and systems. Under these conditions, systems and information that do not need access to the Internet and have comparable security requirements should be grouped within a segregated network. The organisation can repeat this with other groups of related systems and services, which helps to streamline how security controls are applied.

This can be achieved with either separate networks or VLANs as necessary and sensible for the organisation's infrastructure and business processes.

6.3 Implement secure protocols

Insecure protocols should be disabled on all hosts and only the necessary secure protocols used. Many insecure protocols are still in common use, so it's sensible to confirm that all protocols that ransomware could use to move through the network are identified and addressed. For instance, SMBv1 is still commonly used despite being at least partially responsible for the spread of WannaCry in 2017. This should be updated to SMBv2 or SMBv3.

This should be addressed as part of a broader process of auditing the protocols used by and ports open on each host.

Any insecure protocols that are necessary should be clearly documented, along with the justification for using them. Where possible, you should apply additional controls to manage any risks related to using those protocols.

8.4 Limit user permissions to the minimal

The principle of least privilege is well-established good practice. Not only does it protect sensitive information from being accessed accidentally or by nosy users, it also limits ransomware's ability to spread through a network and onto related networks.

Users should only have the privileges they need in order to do their jobs. Any extended privileges should be documented and time-limited so they can be reviewed and removed as necessary.

Where possible, the defined least privileges should be applied on a group level. This makes it simpler to keep track of users with elevated privileges (whether on a temporary or permanent basis) and control changes to what is considered necessary when business processes change.

9.7 Define software restriction policies

Defining a software restriction policy can help prevent malicious code from executing, and can be defined in group policies to ensure they are appropriate to the user. Admins can set up policies to specify which software can execute on clients, prevent specific software from running, specify which users can install software, and/or establish rules by host, local network, domain, etc.

Software restriction policies can be as strict as you need, and defined across the network or on a host-by-host basis to customise restrictions as necessary.

9.9 Whitelist applications

Much like software restriction policies, whitelists can be used to prevent unwanted software and applications from running. Tools like AppLocker and application control can be used to prevent unauthorised applications from running.

Microsoft describes application control as "a crucial line of defense for protecting enterprises given today's threat landscape", and that it "moves away from an application trust model where all applications are assumed trustworthy to one where applications must earn trust in order to run".[18] As such, it provides a good introduction to a zero-trust model, which is an effective method of protecting the organisation from malware.

The zero-trust model is a response to older models that relied on boundary defences. The older model assumes anything outside the network was potentially hostile but failed to address internal risks with the same vigour. Under the zero-trust model, all activities inside and outside the network are considered potentially hostile until shown otherwise.

9.10 Run unknown applications in a sandbox

When assessing any executable application or file, it should be initially only run in a sandbox isolated from any network

---

[18] Microsoft, "Application Control for Windows", May 2020, *https://docs.microsoft.com/en-us/windows/security/threat-protection/windows-defender-application-control/windows-defender-application-control*.

and the Internet. This enables the application or code to be tested without risking any other part of the network.

Sandboxes can be run in virtual environments, so no additional hardware is necessary. The environment you establish in the sandbox doesn't need to be complicated or replicate a normal working environment in most cases.

The Cyber Essentials scheme sets out some conditions for sandboxing. These require that the sandbox prevent access to:

- Other sandboxed applications;
- Data stores;
- Sensitive peripherals; and
- Local networks.

CHAPTER 6: ADVANCED CONTROLS

The following controls extend those set out previously to provide a more advanced, mature anti-ransomware capability. They can be selected on the basis of need, to develop stronger defences or to protect particularly vulnerable or valuable systems and information.

1.4 Incident response training

Incident response is critical for any organisation under attack, but developing the skills relevant for dealing with ransomware can be difficult or prohibitively expensive for smaller organisations. In many cases, this role is best filled by third parties, the cost of which might be covered by insurance.

In addition to the basics of response that should be delivered in staff training as set out in controls 1.1 and 13, organisations may need to develop expertise in isolating infections, tracing movement through the network and identifying the specific ransomware. Where the ransomware can be identified, it's possible that there are solutions that will decrypt data and systems affected by the ransomware. Several websites track known keys for older ransomware.[19]

The response should be led by someone trained and knowledgeable, supported by a team with the skills to complete all of the necessary activities: isolating infected

---

[19] For instance, *www.nomoreransom.org/*.

devices, tracking down any potential sites of infection, identifying the malware (if possible), checking for decryption keys or fixes, cleaning infected devices and securely returning them to operation.

3.3 Maintain blacklists

Given the ease with which new domains can be set up and the need to get around blacklists, criminals will regularly establish new sites from which to launch attacks. It's important that blacklists used to filter malicious content are kept up to date. These are primarily managed through browsers and anti-malware software, and regular updates should be available from vendors.

Many services also provide regular updates to blacklists via intelligence feeds. A sensible approach is to subscribe to a number of these feeds and apply their updates to your blacklists, which can be automated.

4.5 Implement filename spoofing prevention

Filename spoofing is a technique used to disguise executables as other kinds of files, such as PDFs or JPGs. This is generally made more effective by changing the icon to match, which is a trivial task. As such, a user could be presented with a file that uses the text file icon and appears to be called 'file.txt', but is in fact an executable malware file.

Some malware developers will also use right-to-left override (RTLO) to flip the last seven or so characters in a filename, which can make a filename such as malwarefdp.exe appear to be malwareexe.pdf. This can also be done to make files inserted into the registry appear to be legitimate. RTLO is legitimately used to support languages that are read from right to left, and can be triggered using Unicode characters.

On the plus side, this technique does not fool anti-malware and will be identified by most mail applications, but it should be prevented in general by amending Windows' default settings to show extensions for known file types.

5.1 Determine location and RTO of critical data

In order to adequately back up data and prioritise recovery operations, you need to understand the criticality of the data and systems affected. The recovery time objective (RTO) is a measure stating how quickly data needs to be recovered before the impact becomes critical. For instance, if your organisation relies on client contact data and the cost of not having that becomes untenable after two days, then the RTO for client contact data must be less than two days.

It usually isn't necessary to define RTOs for all data – a lot of data simply isn't critical enough to cause issues unless there is significant difficulty recovering from an attack. As such, defining RTOs should begin with an exercise to identify what data is critical and where it is within the organisation's network and systems.

As previously mentioned, the RTO also influences the frequency of backups. In the above example, the organisation needs to ensure that within the RTO it can recover enough data to function, so the backups should be taken more frequently – if ransomware hits right before a backup and it takes nearly two days to recover, the organisation has lost almost four days' data, which is also likely to be disastrous.

Recovery processes will need to take RTOs into account to ensure that backups can be restored within the necessary time frames. Having to recover data held on other media from a remote location may add to difficulties, for instance, not to mention the availability of workstations and other devices that can be deployed while infected machines are being scrubbed. This means that backup storage needs to bow to the need to recover within these timelines.

6.2 Prevent workstation-to-workstation traffic

Aside from some tasks for IT admins, direct communication between workstations is rarely necessary and presents a key vector for ransomware spreading through a network. By default, any such traffic should be blocked, and any exceptions should have additional controls to manage risks. For instance, if direct traffic between two workstations is necessary, they might be placed in a segregated network with no Internet access.

Any traffic that needs to go from one workstation to another should instead be routed via a server with traffic controls to prevent the spread of ransomware and other malware.

6.4 Implement only necessary traffic between network segments

Because network segmentation limits the spread of malware by restricting its access to other devices, connections between different segments must also be strictly controlled. Firewalls on each segment can be configured to control traffic and eliminate it where there is no need for communication.

6.5 Implement MAC/IP anti-spoofing protection

Cyber criminals will frequently attempt to mask their identity (IP and/or MAC addresses) in order to circumvent network defences or redirect users to malicious sites. Attackers can spoof their IP and MAC addresses in a number of ways, so the organisation should take steps to prevent this.

The organisation can prevent these techniques using a variety of tools, such as checking digital certificates to validate signatures.

7.1 Block access to C&C networks

Most modern ransomware relies on communication with a command and control (C&C) server to direct its activities. Blacklisting services maintain lists of known C&C servers, so the organisation can procure regular updates to blacklists and prevent any ransomware from communicating with them.

7.2 Subscribe to intelligence feeds

Keep blacklists of command and control servers up to date by subscribing to security feeds. Just as with websites, it is relatively trivial for criminals to establish new C&C servers. More than that, the criminal organisations that provide these servers have a strong financial motivation to do so – an unknown C&C server can be rented out at a far higher rate than those that feature on a number of blacklists. As such, it's important that firewall blacklists are kept up to date.

Many services also provide regular updates to blacklists via intelligence feeds. A sensible approach is to subscribe to a number of these feeds and apply their updates to your blacklists.

8.3 Implement MFA for admin accounts

To prevent brute-forcing of credentials, multifactor authentication (MFA) should be mandatory for all admin accounts as a minimum. Depending on need, it may be worth rolling out MFA across a broader section of the estate, and establishing more complex MFA requirements for some systems.

9.2 Prevent applications running from user-writable areas of the hard drive

Ransomware (and other malware) obviously needs to execute in order to have an effect. In line with restricting the software that a user can install or execute, preventing applications from running in user-writable areas of the hard drive can prevent this from happening at all.

This implicitly needs to also prevent unknown executables from being saved to areas of the hard drive from which applications can run. The obvious solution is to restrict the user's write access to those areas.

9.8 Disable Windows Script Host and PowerShell

Some malware relies on native applications to provide some of the functionality – in particular, Windows Script Host (WSH) and PowerShell. Because these can execute a range of functions on Windows, they provide an easy way to automate the ransomware's functions. Furthermore, as the instructions that run on WSH and PowerShell are not in executable files, they can often be mistakenly clicked by users who are wary of .exe and .bat, but not aware of the risk in .vbs.

These scripts can also be launched from the startup folder, which eliminates the need to have the user initiate them. Malware that's been removed from a device may have left such files behind, allowing the infection to persist.

If these can't be disabled, the organisation should monitor their use.

10.1 Monitor for known malicious file extensions

While every file type has a legitimate purpose, some are consistent vectors for malware. As such, any external files with the following extensions should be carefully monitored:

- .bat
- .doc/.docm
- .exe
- .lnk
- .pdf
- .pif
- .rtf
- .scr
- .vbs
- .xls/.xlsm
- .zip

While it's probably counterproductive to block these entirely, applying security features (such as Safe Attachments in Office 365 ATP) to scan these files on entry to the organisation is probably not going to reduce productivity in any notable way.

10.2 Monitor for increase in file renames

A lot of ransomware will rename files as it encrypts them, so rapid renaming of files is often a sign that you have an infection. Identifying when ransomware is in action can be an effective backup measure to help you minimise the impact. While it is not preventive, being able to quickly respond to the infection will almost certainly reduce the remediation costs.

There are many tools that will monitor activities such as this, and many of their functions will also identify other malware in action. When this sort of activity is identified, your first step should be to lock down the device and prevent further writing.

10.3 Implement sacrificial network shares/data files

If you suspect you're likely to be subject to ransomware attacks, it can be useful to establish honeypots to identify trends in attack styles, tools and so on. Evidence of being a target might include increasing volumes of phishing emails, consistent patterns of attack, network boundary event logs, and so on. Evidence gathered from the honeypot can be passed to the appropriate authorities.

For obvious reasons, honeypots should be wholly isolated from the rest of the network.

Sacrificial files can also be useful. Fake documents that appear important are likely to attract attention from criminals, so monitoring them for access and changes can provide a valuable early warning of infection or intrusion.

10.4 Implement IPS/IDS with exploit kit detection

An intrusion detection system (IDS) or intrusion prevention system (IPS) should be implemented with exploit kit detection capabilities. IDS and IPS are obviously valuable tools that raise security alerts that can help you either prevent intrusions entirely or mitigate any damage, but it is important to remember that – much like anti-malware software – their engines and signatures need to be kept up to date.

Exploit kits are common tools used to automatically and silently exploit vulnerabilities on the victim's device, usually while they browse the web. They are often used to distribute malware or gain remote access. They are normally used on compromised websites to profile the visitor's browser-based applications and look for vulnerabilities. Because they don't directly attack the network, they can be difficult to spot in action.

10.5 Monitor processes that read or write too many files too quickly

This is closely related to identifying the rapid renaming of files as a symptom of infection. While there are many applications that will rapidly read and/or write files, there should be a limited, known number of such applications on the network. Throttling such activity may be an effective method of slowing the impact of ransomware without accidentally preventing valid business activity.

10.6 Deploy file integrity/file change monitoring software

A change-detection mechanism such as a file-integrity monitoring tool should raise alerts when there are unauthorised changes (including deletion) to a range of file types. This should especially include critical system files and configuration files, but file-monitoring tools can also be configured to pay special attention to key data files that might be subject to attack, such as files containing sensitive or valuable information. The solution should also be configured to compare critical files against previous versions at least weekly.

The solution should be configured to monitor system executables, application executables, configuration and parameter files, log and audit files (including archives), and any other files considered critical.

If the solution is incorrectly implemented or the output isn't monitored, ransomware or an attacker could modify, delete or encrypt files, with an obvious impact on related systems. Undetected changes could also affect security measures, rendering other controls ineffective.

FURTHER READING

IT Governance Publishing (ITGP) is the world's leading publisher for governance and compliance. Our industry-leading pocket guides, books, training resources and toolkits are written by real-world practitioners and thought leaders. They are used globally by audiences of all levels, from students to C-suite executives.

Our high-quality publications cover all IT governance, risk and compliance frameworks and are available in a range of formats. This ensures our customers can access the information they need in the way they need it.

Our other publications about cyber security include:

- *The Cyber Security Handbook – Prepare for, respond to and recover from cyber attacks with the IT Governance Cyber Resilience Framework (CRF)* by Alan Calder, *www.itgovernancepublishing.co.uk/product/the-cyber-security-handbook-prepare-for-respond-to-and-recover-from-cyber-attacks*
- *Securing Cloud Services – A pragmatic guide, second edition* by Lee Newcombe, *www.itgovernancepublishing.co.uk/product/securing-cloud-services-a-pragmatic-guide*
- *The Psychology of Information Security – Resolving conflicts between security compliance and human behaviour* by Leron Zinatullin,

*www.itgovernancepublishing.co.uk/product/the-
psychology-of-information-security*

For more information on ITGP and branded publishing services, and to view our full list of publications, visit *www.itgovernancepublishing.co.uk*.

To receive regular updates from ITGP, including information on new publications in your area(s) of interest, sign up for our newsletter at *www.itgovernancepublishing.co.uk/topic/newsletter*.

Branded publishing

Through our branded publishing service, you can customise ITGP publications with your company's branding.

Find out more at:
*www.itgovernancepublishing.co.uk/topic/branded-
publishing-services*.

Related services

ITGP is part of GRC International Group, which offers a comprehensive range of complementary products and services to help organisations meet their objectives.

For a full range of resources on cyber security visit *www.itgovernance.co.uk/cyber-security-solutions*.

Training services

The IT Governance training programme is built on our extensive practical experience designing and implementing management systems based on ISO standards, best practice and regulations.

Our courses help attendees develop practical skills and comply with contractual and regulatory requirements. They also support career development via recognised qualifications.

Learn more about our training courses in cyber security and view the full course catalogue at *www.itgovernance.co.uk/training*.

Professional services and consultancy

We are a leading global consultancy of IT governance, risk management and compliance solutions. We advise businesses around the world on their most critical issues and present cost-saving and risk-reducing solutions based on international best practice and frameworks.

We offer a wide range of delivery methods to suit all budgets, timescales and preferred project approaches.

Find out how our consultancy services can help your organisation at *www.itgovernance.co.uk/consulting*.

Industry news

Want to stay up to date with the latest developments and resources in the IT governance and compliance market? Subscribe to our Weekly Round-up newsletter and we will send you mobile-friendly emails with fresh news and features about your preferred areas of interest, as well as unmissable offers and free resources to help you successfully start your projects. *www.itgovernance.co.uk/weekly-round-up*.

EU for product safety is Stephen Evans, The Mill Enterprise Hub, Stagreenan, Drogheda, Co. Louth, A92 CD3D, Ireland. (servicecentre@itgovernance.eu)

www.ingramcontent.com/pod-product-compliance
Lightning Source LLC
Chambersburg PA
CBHW070850070326
40690CB00009B/1784